24th Topical Conference on Gas Utilization 2024

Topical Conference at the 2024 AIChE Spring Meeting and 20th Global Congress on Process Safety

New Orleans, Louisiana, USA
24-28 March 2024

ISBN: 979-8-3313-1708-9

24th Topical Conference on Gas Utilization 2024

Topical Conference at the 2024 AIChE Spring Meeting and 20th Global Congress on Process Safety

New Orleans, Louisiana, USA
24-28 March 2024

AIChE
The Global Home of Chemical Engineers

Printed from e-media with permission by:

Curran Associates, Inc.
57 Morehouse Lane
Red Hook, NY 12571

Some format issues inherent in the e-media version may also appear in this print version.

Printed with permission by Curran Associates, Inc. (2025)

For permission requests, please contact AIChE
at the address below.

AIChE
120 Wall Street, FL 23
New York, NY 10005-4020

Phone: (800) 242-4363
Fax: (203) 775-5177

www.aiche.org

Additional copies of this publication are available from:

Curran Associates, Inc.
57 Morehouse Lane
Red Hook, NY 12571 USA
Phone: 845-758-0400
Fax: 845-758-2633
Email: curran@proceedings.com
Web: www.proceedings.com

TABLE OF CONTENTS

DECARBONIZATION OF GAS PROCESSING

GAS UTILIZATION (METHANE, ETHANE, AND CO2)

KEYNOTE - 22ND TOPICAL CONFERENCE ON GAS UTILIZATION

LNG AND HYDROGEN SAFETY

GAS UTILIZATION (METHANE, ETHANE, AND CO2) II

DECARBONIZATION OF GAS PROCESSING II

77b Investigating the Effect of Dehydrating Agents for the Direct Synthesis of Dimethyl Carbonate and Study Their Potential for Carbon Dioxide Fixation

Amera Abdelbar, Hanif Choudhury, Mohamed Sufiyan Challiwala, Gasim Ibrahim, Nimir Elbashir

Dimethyl carbonate (DMC) is known to be a safe and non-corrosive material that has wide range of applications. For example, it can be used as a fuel additive for gasoline and diesel to improve their combustion efficiency and reduce the pollutants emission. DMC has been manufactured through different methods such as the phosgenation of methanol and methanol oxidative carbonylation. However, these routes involve the use of toxic and hazardous materials such as carbon monoxide (CO) and phosgene. Therefore, there is growing interest in alternative routes, such as direct synthesis from carbon dioxide (CO2) and methanol (MeOH). This method is considered to be an attractive alternative, due to its low toxicity compared to the other routes. Moreover, it offers environmental benefits by utilizing CO2, thereby contributing to the reduction of atmospheric CO2 levels. According to different studies reported in the literature, DMC is considered one of the most promising products derived from CO2 utilization and it has the potential to serve as a model reaction for Carbon Capture and Utilization (CCU). Furthermore, this method does not include the formation of any toxic or corrosive materials, as water is the only by-product. However, the direct synthesis method faces several drawbacks that limit the methanol conversion towards DMC to around 1% even under favoured conditions of production. Some of these drawbacks are the high stability and low reactivity of CO2, in addition to the in-situ formed water, as a by-product, that shifts the reaction equilibrium toward the hydrolysis of DMC [1].

Therefore, to overcome the energy barrier of the reaction, an effective catalyst is required. Various catalysts have been investigated and reported in the literature, such as cerium oxide (CeO2), silicon dioxide (SiO2), and zinc oxide (ZnO), among which CeO2 exhibits the most effective performance. Simultaneously, a method for in-situ water removal is required to shift the equilibrium towards DMC production. Different methods have been suggested for in-situ water removal such as the use of dehydrating agents such as 2- cyanopyridine (2-CP), acetonitrile, and benzonitrile. These dehydrating agents reacts with water helping in shifting the reaction towards the DMC formation based on Le Châtelier's principle. 2-CP is the dehydrating agent with the highest performance as it helped in increasing the conversion above 95% with a DMC selectivity greater than to 99% [2]. However, the use of the other nitriles did not increase the conversion as 2-CP and their selectivities were lower probably due to the side reactions such as reaction of the nitriles with methanol, decomposition of the nitriles, and catalyst deactivation. For example, in the case of using benzonitrile as a dehydrating agent, it helped to overcome the equilibrium limitation and to increase the DMC conversion to 47% by reacting with water resulting in the formation of benzamide. However, the DMC formation reaction stopped at this percentage and did not go over it due to benzamide as a catalyst poison [3]. 2,2-dimethoxypropane (DMP) is another dehydrating agent that was reported in literature due to its rapid and effective action by chemical hydrolysis. However, the use of DMP reduced the DMC selectivity due to the formation of acetone as a by-product. Despite the outstanding performance of 2-CP, it still did not help in scaling up the process and making it economically feasible due to difficulties in separation and regeneration. Therefore, the search of other dehydrating agents continuous. In this work, dehydrating agents molecular such as zeolite 3A, 4A, 5A were tested due to their large surface area, thermal stability, and excellent adsorption. However, the addition of large amount of molecular sieve may lead to reduction in the catalyst activity as it can prevent the interaction of reactants with the catalyst active sites. On the other hand, the use of small amounts of molecular sieve will not be to remove the formed water and achieve optimal conversion [4]. Moreover, ethylene glycol was also tested as it showed high potential as a candidate due to its utilization in the natural gas industry to dehydrate the natural gas.

This work also aims to investigate the ability of the DMC process in the presence of different dehydrating as a CCU reaction pathway to fix CO2. This goal is achieved by using the CO2Fix parameter, which is a matric developed by Ibrahim et. al. to estimate the CO2 fixation potential for different CCU pathways. If the CO2Fix calculations resulted in a value equal to or above 1, then the process does utilize more CO2 than it produces, however, a CO2Fix value less than 1 indicate that the process produces more CO2 than it utilizes [5]. For the direct synthesis of DMC from CO2 in the presence of dehydrating agents such as 2-CP, acetonitrile, and benzonitrile helped in achieving a conversion equal to 94%, 9%, and 47%, respectively. The corresponding CO2Fix calculations were 3.68 for 2-CP, 2.92 for benzonitrile, and 0.80 for acetonitrile. Meaning that the direct synthesis method has the potential to fix CO2 in the presence of 2-CP and benzonitrile as dehydrating agents, but not in the presence acetonitrile. Furthermore, within this work, the overall direct DMC synthesis process shall be simulated to study the economic aspects.

References

1. Pyo, S.-H., Park, J. H., Chang, T.-S., & Hatti-Kaul, R. (2017a). Dimethyl carbonate as a green chemical. Current Opinion in Green and Sustainable Chemistry, 5, 61–66. https://doi.org/10.1016/j.cogsc.2017.03.012

2. Honda, M., Tamura, M., Nakagawa, Y., Nakao, K., Suzuki, K., & Tomishige, K. (2014). Organic carbonate synthesis from CO2 and alcohol over CEO2 with 2-cyanopyridine: Scope and mechanistic studies. Journal of Catalysis, 318, 95–107. https://doi.org/10.1016/j.jcat.2014.07.022

3. Tomishige, K., Tamura, M., & Nakagawa, Y. (2018). Co2 conversion with alcohols and amines into carbonates, ureas, and carbamates over CEO2 catalyst in the presence and absence of 2-cyanopyridine. The Chemical Record, 19(7), 1354–1379. https://doi.org/10.1002/tcr.201800117

4. Faria, D. J., Moreira dos Santos, L., Bernard, F. L., Selbacch Pinto, I., Carmona da Motta Resende, M. A., & Einloft, S. (2020). Dehydrating agent effect on the synthesis of dimethyl carbonate (DMC) directly from methanol and carbon dioxide. RSC Advances, 10(57), 34895–34902. https://doi.org/10.1039/d0ra06034h

5. Ibrahim, G., Challiwala, M. S., Choudhury, H. A., Zang, G., El-Halwagi, M. M., & Elbashir, N. O. (2023). CO2Fix: An approach to assess CO2 fixation potential of CCU reaction pathways. Computers & Chemical Engineering, 178, 108398. https://doi.org/10.1016/j.compchemeng.2023.108398

77c Application of Greenhouse Gas Abatement, Sustainability and Economic Framework (GASEF) Tool for Combined Evaluation of Economic and CO_2 Fixation Potential of Ccu Reactions

Gasim Ibrahim, Mohamed Sufiyan Challiwala, Mahmoud El-Halwagi, Nimir Elbashir

A global endeavor is in progress to diminish greenhouse gas (GHG) emissions in response to mounting concerns regarding human-induced climate change. In 2020, industrial greenhouse gas emissions amounted to 8.5 $GtCO_2$, contributing to roughly 24% of total global anthropogenic emissions. Consequently, the industrial sector must cut GHG emissions by 1.2% to reach 7.4 $GtCO_2$ by 2030 to fulfill sustainability and decarbonization objectives. The utilization of carbon dioxide capture and utilization (CCU) has received significant attention for its potential to reduce GHG emissions and decarbonize several industries, such as the natural gas, cement, and steel manufacturing sectors. One way to integrate CCU into a process is by chemically converting CO_2 into alternative products. It's worth noting that not all CCU processes may be economically feasible or environmentally sustainable with regard to CO_2 capture. Typically, assessing the CO_2 capture of a specific CCU reaction involves conducting a life cycle assessment (LCA), and the economic viability of the process is evaluated through a technoeconomic assessment (TEA). These assessments necessitate a substantial amount of theoretical and empirical data, as well as a significant investment of time and effort. Consequently, it is advisable to conduct these assessments in a later stage of the development and scaling up of CCU technology.

In this work we develop a tool based on the CO2Fix metric and the GASEF to evaluate the economic and CO_2 fixation potential of CCU reactions. The CO2Fix provides an estimate for the CO_2 fixation potential of a CCU reaction based on limited reaction data. The GASEF combines this with the Metric for Inspecting Sales and Reactants (MISR) which estimates economic potential, providing a combined economic and CO_2 fixation analysis. This framework classifies CCU conversion technologies into four types: Type 1 is both economically feasible and leads to CO_2 fixation, Type 2 leads to CO_2 fixation but is not economically feasible, Type 3 is economically feasible but does not lead to CO_2 fixation, and Type 4 is neither economically viable nor leads to CO_2 fixation. Additionally, this framework allows the estimation of various parameters, such as minimum subsidies needed for Type 2 technologies, minimum CO_2 capture for Type 3 technologies to achieve CO_2 fixation, minimum CO_2 credits for Type 2 technologies, and maximum CO_2 capture costs for Type 3 technologies while maintaining economic viability.

The tool is used to evaluate common CCU reactions such as the dry reforming of methane (DRM), the reverse water gas shift reaction (RWGS), CO_2 hydrogenation to methanol (CO_2 to methanol), direct synthesis of DMC, and the CARGEN reaction. For each technology the tool is used to estimate the CO_2 credits or CO_2 capture and storage cost required for commercial viability. Additionally, the tool is used to compare the technologies in terms of CO_2 fixation efficiency, and their economic potential when economic incentives from CO_2 fixation are taken into consideration.

117a Enhanced Gas Utilization at Pilot Scales

Veshal Venkat, Eloy Flores III, Michael Hartmann

At SwRI, we engage in the process development and improvement for a variety of sustainable fuels and chemicals from lab to pilot scales. The applied research, development, testing and evaluation methods undertaken help bridge the gap between fundamental science and commercialization. The technologies developed include but are not limited to hydrogen production, CCU, renewable fuels, chemicals and materials production and conventional thermo, photo and electro-chemical conversions. Over the years, immense knowledge has been gained from the design, build-up and operations of such processes at various scales. Many limitations such as feedstock preprocessing and feeding, reactor optimization, downstream processing as well as troubleshooting of pilot scale systems have been overcome and understood. The systems have been developed to ensure process safety, enhanced conversion efficiencies, improved energy and mass transfer as well as overall emissions reductions and carbon neutrality. One such system is the production of turquoise hydrogen and value added carbon from natural gas. The current pilot scales systems have been engineered for improved hydrogen production in circulating fluidized bed reactors. The system is continually being screened for process improvement to bring down hydrogen costs below \$2/kg. Other gas conversion systems developed include CO2 conversion to olefins and alcohols, and oxidative coupling reactions for ethylene through novel catalyst and reactor systems. This presentation will include the different approaches to the utilization of value addition of various gases.

117b Catalytic Conversion of CO_2 to Syngas through Methane Dry Reforming and Reverse Water Gas Shift Reactions

Parisa Ebrahimi, Mohammed Al-Marri, Majeda Khraisheh, Anand Kumar

CO_2 chemical conversion into value-added chemicals and fuels, in addition to reducing greenhouse gas emissions has significant economic benefits. Among the most promising methods of converting CO_2 into CO are the Methane Dry Reforming (MDR) and the Reverse Water Gas Shift (RWGS) reactions. By adding an additional unit, such as the Fischer-Tropsch process, utilizing CO and H_2 (syngas) as feedlines, both of these reactions (MDR and RWGS) can be used to generate fuels and a wide range of hydrocarbons. The MDR process has re-emerged as one of the most desirable techniques for the production of syngas in light of its compelling economic and environmental advantages in recent years. These reactions have yet to be commercialized due to the rapid deactivation of the catalysts and the deposition of carbon species on the catalysts. Thus, due to the fact that catalyst choice can have a significant impact on reaction performance, developing catalysts that are highly stable and ensuring that they lead to efficient CO production is critical and challenging. Specifically, this study aims to evaluate the performance of transition metals (such as Ni, Co, and Cu) in RWGS and MDR reactions when supported on commonly used oxides (such as CeO_2, SiO_2, ZrO_2). The results (to be presented) do illustrate the selective nature of Cu metal on CeO2 support, resulting in high activity, CO selectivity, and carbon tolerance over a long period of time on-stream (TOS).

117c Data-Driven Decision-Making for Flexible Natural Gas Allocation Under Uncertainties

Noor Yusuf, Ahmed AlNouss, Tareq Al-Ansari

The global growth in population imposes greater demand on energy resources and synthesised products. In addition to applications for power generation, natural gas is a fundamental feedstock to different industries for producing chemicals and consumer products. Hence, a producer must optimise natural gas monetisation to different processes in alignment with the market needs. Optimal allocation does not only support maximising profitability, but targets such as risk management and environmental mitigation could be satisfied. For the state of Qatar, the natural gas industry is the country's most active and bullish sector. Industries such as liquified natural gas (LNG), gas-to-liquids (GTL), methanol, ammonia, urea, and ethylene dominate the sector for cost-effective natural gas monetisation. However, the increased uncertainties in final markets induced by shocks and environmental regulations have impacted the sellers significantly in the last few years. In this context, this study investigates the economic profitability of flexible natural gas allocations to different utilisation routes based on market data. First, process data were obtained from simulating the processes using Aspen HYSYS. Second, simulation results, forecasted annual price and demand data, and operational production bounds were used as inputs to an agent-based model for deciding on optimal yearly allocation. Assessment of a Qatar-based case study revealed that power and LNG are fellfield, while an additional profit was obtained from methanol production. Moreover, considering environmental constraints, urea production was identified to offset environmental emissions. The results conclude that flexible natural gas allocation to utilisation routes allows the decision-maker to respond proactively to market changes. In alignment with the "real options theory", the approach gives the producer the right, but not the obligation, to switch production capacities to seize market opportunities when the market is bullish and minimise risks when the market is low.

117d Catalytic Bubble Column Reactors with Simultaneous C-H Bond Activation, C-O Bond Activation, Solid Carbon Co-Production, and Carbon Separation

Chester Upham

Production of hydrogen via methane pyrolysis has gained interest in recent years as new reactor concepts have been reported. Bubble column reactors allow for continuous separation of carbon co-product. Recently, we have reported catalytic melts that simultaneously active C-H bonds and the C-O bonds in CO_2[1,2]. The resulting products are syngas and solid carbon. We have tested over 50 metals and metal alloys, and activity trends for some of them will be discussed. Indium and indium alloys activate CO_2 at the highest rate, and Sn-In alloys balance C-O and C-H bond activation most effectively. Technoeconomic considerations will also be discussed based on our recent work[3,4].

[1] Cai, Zong, Smith, and Upham, *Chemical Communications*, Issue 52, 2023

[2] Palmer, Upham, Smart, Gordon, Metiu, and McFarland, *Nature Catalysis*, 3 (1), 2020

[3] Zong, Cai, Tabbara, and Upham, *Energy Conversion and Management*, 277, 2023

[4] Zong, Koers, Cai, and Upham, *in review*, 2023

31a CO_2 Utilization Technologies to Support Decarbonization Journey

Lina Rueda

There is a dire need for technological innovation and disruption to meet the increasing demand for cleaner energy mix in the world and for circumventing CO_2 emission impacts. This will help to meet the "Net Zero" target for the scope 1, 2 and 3 emissions by 2050, but also create a "Positive Impact' by aiming for negative emission by targeting historical emissions through the technologies like Direct Air Capture (DAC). At Qatar Shell Research and Technology Center (QSRTC), we are working on developing such technologies, in particular to support the decarbonization journey which remains a gigantic task to achieve reduction in emissions from "35Gt CO_2e to Net Zero".

Besides, CCS & DAC, the CO_2 valorization will play an important role in decarbonization journey, as it creates new value chains and fulfills the broader goal of circularity under the sustainability vision. In fact the use of CO_2 to create valuable products can contribute to reducing the high cost associated with CCS (carbon capture and storage) and support the acceleration of the decarbonization journey. This presentation will provide a high-level overview of on-going technology developments at QSRTC at different stages:

- Cargen: Co_2 Utilization for The Co-Production of Carbon Nanotubes & Syngas
- Sustainable Conversion of CO_2 to Organic Carbonates
- An Integrated Process for The Simultaneous Desalination of Reject Brine and CO_2 Mineralization
- Semi-Continuous Reactor Engine
- Electrochemical H_2 Production from Seawater using Membraneless Electrolyzer Coupled with CO_2 Utilization

31b Kinetics Model Development and Experimental Scale-up of Cargen® Technology from Milligram Scale to Multi-Kilogram Scale

Nimir Elbashir, Mohamed Sufiyan Challiwala, Hanif Choudhury, Gasim Ibrahim

Dry reforming of methane (DRM) offers an avenue for converting carbon-dioxide (CO_2) and methane (CH_4), the two major greenhouse-gases, into syngas— a vital chemical precursor. However, DRM is constrained by high energy-demands, catalyst-deactivation, and an unfavorable H_2/CO ratio. Previously, a unique dual-reactor system that produces multi-walled carbon nanotubes (MWCNTs) and syngas as products was proposed that offers at least 65% CO_2 conversion at 50% the energy demands of DRM. The present study experimentally proves and scales the concept from milligram scale to multi-gram scale and finally to multi-kilogram scale of MWCNT production. This study also introduces and experimentally validates a lumped Langmuir-Hinshelwood-Haugen-Watson (LHHW) kinetics model capturing a network of nine-primary-reactions involving CO, O_2, H_2, CH_4, CO_2, H_2O, and solid carbon. At milligram scale, the model performs within 5% error-margin for carbon-formation rate at 550°C. At multi-gram scale, the model is validated at 500, 550, and 600°C to capture temperature effect. At this scale, the CH_4-conversion and CO2-conversion predictability is within 8% and 30% error-margin respectively. At multi-kilogram scale, the model predicts carbon-formation rate within 21% error-margin at 550°C. Finally, characterization of the MWCNTs using Raman, SEM, TEM, STEM, and TGA-DTA confirms MWCNT quality consistency at all the scales.

In this presentation, we will highlight the unique journey of the scale-up of the CARGEN® process from milligram-scale to multi-kilogram scale demonstrating its readiness for commercial implementation for the large-scale decarbonization and bulk MWCNT production.

SPRING24 +20TH GCPS
A Joint AIChE and CCPS Meeting

20th Global Congress on Process Safety
March 24-28, 2024 | New Orleans, Louisiana

Evaluation Protocol for Vapor Cloud Explosion Models

Filippo Gavelli (fgavelli@BlueEandC.com)
Thomas Butts (tbutts@BlueEandC.com)
Blue Engineering and Consulting Company

Abstract

The siting of LNG facilities requires several different hazards to be quantified in order to properly assess the risk to the public and public property in areas surrounding a facility. For each type of hazard, different types of modeling tools are generally available, ranging from empirical models to 3D computational fluid dynamics (CFD) packages; this complicates the evaluation of modeling results by the regulatory agency, which may not have the expertise to assess the accuracy of each model. To fill this gap, a project was initiated to develop a set of model evaluation protocols for each major type of hazard associated with LNG facilities.

This paper describes the Model Evaluation Protocol (MEP) developed for Vapor Cloud Explosion (VCE) models, specifically as applied to LNG facility siting studies. The paper will discuss the evaluation methodology, including the information to be submitted by the model proponents and the review process, the data set to be used for model validation, and the criteria to evaluate each model and to define their respective validation factors (i.e., safety factors).

Keywords: model evaluation; model validation; VCE; overpressure; LNG

1 Introduction

The safe siting of LNG facilities requires project developers to evaluate the potential consequences to the public and public property from accidental loss of containment and release of hazardous materials (e.g., flammable and/or toxic). These consequences include: pool, jet or flash fires; vapor cloud explosions; toxic gas exposure; and catastrophic vessel failures. Calculating the distance to which hazardous conditions may extend for each type of scenario, to the thresholds specified by regulators, requires the use of computational models.

Assessing the quality of consequence models used to determine public safety has long been a concern to regulatory bodies. Therefore, the Pipeline and Hazardous Materials Safety Administration (PHMSA) commissioned Blue Engineering and Consulting Company (BLUE) and GTI Energy (GTI) to develop a set of MEPs, that will allow the review of modeling tools for several of the hazards listed above.

The scope of the MEP presented in this paper is to evaluate models that calculate the overpressure hazards generated from the ignition of a flammable vapor cloud – i.e., a vapor cloud explosion. The complete documents (respectively, the model evaluation protocol and the associated model validation database for VCE models can be obtained from PHMSA's project website at: https://primis.phmsa.dot.gov/matrix/PrjHome.rdm?prj=854.

1.1 Scope of the Model Evaluation

The scope of the VCE MEP is to define an objective and transparent methodology to determine the ability of a model to accurately quantify overpressures for vapor cloud explosion scenarios typically evaluated in LNG facility siting. The protocol is intended to be a guidance document which outlines the tasks required by the petitioner and the reviewer. The objective of evaluation is to allow the regulatory body (i.e., PHMSA) to determine whether a given model should be approved or rejected for use in VCE modeling within the context of LNG facility siting studies. Additionally, the evaluation should allow the regulatory body to specify any restrictions on the use of an approved model, as well as any "validation factors" to be applied to the modeling results. It is important to note that comparing models against one another or establishing rankings of preferred models are not objectives of the MEP.

1.2 Vapor Cloud Explosions

In order for a VCE to occur, several previous steps are required in the hazard sequence: the loss of containment of a flammable inventory needs to occur; the released stream must form a cloud which is not immediately ignited; the flammable cloud grows and/or migrates to a confined and/or congested region, at which point it is ignited. This simplified description of the sequence of events leading up to a VCE is sufficient to demonstrate that the outcome of a VCE is affected by a large number of parameters, some of which are briefly described below.

- Fuel: Several different flammable materials may be present in LNG facilities, for the most part hydrocarbons or hydrocarbon mixtures. While each flammable material or mixture has specific thermophysical properties (density, boiling point, enthalpy, etc.), the main distinction between materials with respect to VCE modeling is associated with their reactivity. On the basis of reactivity, flammable materials in LNG facilities are typically grouped as follows:

- LNG and natural gas are composed primarily of methane and, therefore, have low reactivity. As a result, methane mixtures are less likely to produce damaging overpressures than other fuels, all other conditions being equal.
- Ethylene is a refrigerant component present in many LNG facilities. Ethylene is a high reactivity fuel with a laminar flame speed nearly double that of methane. Consequently, ethylene is more likely to cause damaging overpressures relative to other hydrocarbons.
- Other flammable materials that may be present in LNG facilities include refrigerant components (e.g., propane, pentane, etc.) and mixtures thereof, as well as pretreatment byproducts (e.g., heavy hydrocarbon condensate). These materials generally fall into the category of "medium reactivity" fuels (i.e., less reactive than ethylene but more reactive than methane).

- Flammable Cloud Volume: It is well established that, other conditions being equal, the consequences (i.e., overpressure) from a VCE increase with the size of the flammable cloud inside confined and/or congested region. The volume of a flammable cloud depends on several factors, including: the type of release (e.g., a high-momentum, jet release tends to increase mixing with air, leading to a large, well-mixed cloud, whereas vapors from a liquid spill mix more slowly, keeping the flammable volume smaller); release rate (a larger release will tend to form a larger cloud, however, depending on ventilation and confinement, a large portion of the cloud may remain too rich to burn); ambient conditions, particularly wind speed and wind direction relative to the release; and time before ignition. For LNG facility siting studies, it is usually necessary to determine the worst-case flammable cloud for VCE consequences; in order to simplify this evaluation, Equivalent Stoichiometric Cloud concepts are often applied [1], which convert a non-uniform flammable cloud to an equivalent, uniform stoichiometric cloud whose ignition would yield similar overpressures.
- Concentration Profile: The concentration profile within the cloud can affect how the flame propagates, as near-stoichiometric regions will burn more rapidly than rich or lean regions.
- Congestion: Flow turbulence is a primary mechanism for flame acceleration (and corresponding pressure increase) during a VCE. Turbulence is generated any time a flame front interacts with an obstacle, therefore, the more obstacles (i.e., congestion) are present within an ignited flammable cloud, the higher flame speed and overpressure can be expected (in reality, the size and orientation of the obstacles also affects the turbulence generation).
- Confinement: Any combination of walls, barriers, or large obstacles that prevent the flow or expansion of gas in a specific direction is defined as confinement. In general, confinement leads to the buildup of overpressures because combustion products are unable to freely expand in one or more directions, therefore, the flow velocity increases in the unconfined directions.
- Ignition location: The location of ignition within a flammable cloud can significantly affect the outcome of a VCE, both in terms of maximum overpressure as well as direction of blast propagation. In general, the longer the path through congestion, the faster the flame will

travel, therefore, ignition near the edge of a cloud will tend to yield higher overpressures than ignition near the center of the cloud.

1.3 VCE Models

There are numerous models available to calculate VCE hazards for LNG facility siting, both commercially distributed and freely available. Each model has specific capabilities, solution methods, etc. For the most part, current models fall within one of the following categories:

- TNT Equivalency. The TNT equivalency method is the simplest method for estimating the hazards resulting from a VCE [2]. It works by first determining the chemical potential energy contained within a vapor cloud, then calculating the mass of trinitrotoluene (TNT) that contains the same level of energy as the vapor cloud. Finally, the overpressure for the VCE is estimated by calculating the overpressure generated from the equivalent mass of TNT. This method operates on the assumption that the vapor cloud will explode in the same way as TNT, even though the blast characteristics of high explosives are fundamentally different from those of gas explosions. The overpressure curve typical of TNT has been found to overestimate overpressures near the source while underpredicting them in the far field. Attempts to compensate for the inaccuracies of the TNT equivalency method use an efficiency factor (typically 1-15%), but the results from these calculations are typically too inaccurate to be considered useful. Consequently, the TNT equivalency method is no longer accepted by federal LNG regulatory bodies.
- TNO Multi-Energy. The TNO Multi-Energy method (MEM) is a blast prediction model, which recognizes that the outcome of a VCE do not depend solely on the type and mass of fuel involved in the explosion, but rather on other factors such as the congestion and confinement of the surrounding area [2]; in particular, the MEM is focused on the concept of partial confinement and its effect on blast overpressures. The basic methodology of the MEM involves the identification of partially confined volumes and their relative degree of confinement, then the use of semi-empirical correlations to determine the overpressures generated. The blast wave properties (peak static overpressure, peak dynamic pressure, and impulse duration) are represented as functions of distance from the source of the explosion (all variables are expressed in dimensionless form), for a series of 10 curves based on the initial strength of the blast. Like other 1-dimensional models, the MEM is efficient and fast but has limitations. For example, it assumes pressure to be uniform and equal to the source pressure inside the source region; therefore, it is generally not applicable to VCE consequence modeling within or very close to a congested area.
- Baker-Strehlow-Tang. The Baker-Strehlow-Tang (BST) method is another 1-dimensional model that correlates blast overpressure with distance from the source of the blast, by taking into account not only the type and mass of fuel, but also external factors such as confinement and congestion [3], [4]. The BST method uses curves based on the assumed maximum flame speed for each scenario. Curves are provided for a discrete set of flame speed values, based on the following parameters:
    - Fuel reactivity
    - Degree of confinement of the system
    - Obstacle density

Like other 1-dimensional models, the BST method is efficient and fast but also has its limitations. The BST also assumes the pressure to be uniform and equal to the source pressure inside the cloud; therefore, this method is generally not applicable to VCE consequence modeling within a PES.

- Computational Fluid Dynamics (CFD). CFD models solve the time-dependent Navier-Stokes equations over a discretized, three-dimensional domain. In the most common commercial CFD packages, which implement the Reynolds-Averaged Navier Stokes (RANS) approach, the equations of fluid flow are coupled with turbulence closure models and other equations (e.g., conservation of species for mixtures, heat transfer to/from solid boundaries, etc.) as dictated by each specific scenario. Assuming the resolution of the discretized domain is adequate, CFD models provide the most detailed solution of a VCE scenario among all model types. However, the price for this level of detail comes with the cost of the computing time and resources necessary to perform the simulation. In the context of VCE modeling, the two most important categories of CFD models are:
  - Fixed Flame Speed CFD Models: these models assume a constant flame speed throughout the simulation, unaffected by turbulence generated at obstacles. Fixed flame speed CFD models represent an improvement over 1-D models (e.g., MEM and BST) because they model the interaction of blast waves with obstacles and obstructions; however, they require similar assumptions to those models regarding the maximum flame speed for a given scenario. Therefore, their accuracy hinges upon a proper selection of the flame speed.
  - Variable Flame Speed CFD Models: these models account for the flame-acceleration due to turbulence generation at obstacles. Since flame acceleration is calculated during the simulation, more accurate results can be obtained, particularly within congested and/or confined areas where flame acceleration occurs. However, these models require the most time and resources to perform.

2 The VCE Model Evaluation Protocol

The scope of the VCE MEP is to define an objective and transparent methodology to assess the ability of a model to accurately quantify vapor cloud explosion hazards in the context of LNG facility siting. The MEP is a guidance document which outlines the tasks required by the model petitioner and the reviewer. Specifically, it provides guidance to the petitioner regarding:

- Information to be submitted (e.g., user guide, technical references, etc.);
- Scenarios to be simulated for validation (including any sensitivity analysis); and
- Feedback that should be expected once the evaluation is completed (e.g., whether the model is accepted or not, restrictions on the applicability of the model, or validation/safety factors to be applied to the modeling results).

The protocol also provides guidance to the evaluator, including:

- Requirements that an evaluator must meet to be qualified;
- Tasks to be performed during the evaluation; and
- Reporting requirements.

The VCE MEP follows the general model evaluation methodology described by Gavelli et al. [5] and includes the following tasks:

1. Scope of the evaluation
2. Evaluator qualifications
3. Model description
4. Scientific assessment
5. User-oriented assessment
6. Verification
7. Validation, which includes:
    a. Model Validation Database, with:
        i. Datasets for model validation
        ii. Key variables and physical comparison parameters
        iii. Model performance measures
    b. Sensitivity analysis
    c. Qualitative performance assessment
    d. Quantitative performance assessment
8. Model Evaluation Report, which includes:
    a. Summary of the evaluation (i.e., steps 4 through 7 above)
    b. Best practice guidance
    c. Recommendations for improvement

Note that the model validation (step 7 above) is included in a separate document – the Model Validation Database (MVD). This allows the validation data sets, model performance measures and acceptability criteria to be updated as more information becomes available (e.g., new experimental data sets or results from model evaluations) without requiring a complete revision of the model evaluation protocol. Petitioners should always verify that they are validating their models against the most recent version of the MVD.

3 The VCE Model Validation Database

Model validation is the process of comparing the predictions of a model against experimental data relevant to the types of scenarios that the model is intended to predict. In the context of a model evaluation protocol to be used for regulatory purposes, the term "validation" is not intended in absolute terms but, rather, it implies "relative to the scenarios against which it was compared". In addition to specifying data sets for model validation, the MVD needs to include the information necessary to set up and perform the simulations (i.e., geometry, initial and boundary conditions, ignition location, etc.), as well as define the criteria for comparison of the model predictions with the observations.

The following criteria were applied to datasets reviewed for inclusion in the VCE MVD:

- The fuel and test configuration should be representative of realistic scenarios. The scale of the experiment should also be comparable to actual scenarios.
- The tests should be described in sufficient detail for modelers to set up their simulations in an accurate and consistent manner, with as few assumptions as possible.

- Sensor data should include measurement of the key variables (i.e., pressure). The measurement frequency should be sufficient to accurately capture the blast wave.
- Measurements should be of a sufficient quantity to allow assessment of model performance as well as spotting of modeling trends (e.g., progressive underprediction with increasing distance from the blast source).
- Uncertainty of all measured and derived quantities should be provided, together with a description of the method used to define such uncertainty.
- The data must be available for public dissemination.

In order to optimize the database, the test series were selected to span a broad range of conditions and to strike a balance between the number of trials included for each type of scenario. For that purpose, the test series were grouped according to several parameters as follows:

- Material:
  - Methane
  - Alkane (e.g., propane, butane, etc.)
  - Ethylene
- Congested Region:
  - Unconfined
  - Confined (e.g., inside a building)
  - Complex Geometry (i.e., irregular distribution of congestion)
- Target Location
  - Near field
  - Far field

Numerous test series were reviewed for the VCE MVD; each series was "rated" on a 5-point scale according to several categories, following an approach first introduced by Skjold et al. [6] and applied to previous MEPs ([7], [8]). Table 1 lists the test series included in the database. The need for multiple experimental data sets is due to the multiple parameters which can affect the outcome of a VCE, as previously described. A summary of each test series included in the MVD can be found in the VCE MVD report on PHMSA's project website.

Table 1. Experimental test series included in the VCE MVD.

Series	Fluid	Congested Region	Target Location
SOLVEX	Methane, Propane	Confined	Near field
Troll Process Module	Propane	Complex Geometry	Near field
CMI Offshore Compressor (M24) Module	Methane, Propane	Complex Geometry	Near field
CMI Offshore Separator (M25) Module	Methane, Propane	Complex Geometry	Near field
DNV Module	Propane	Confined	Near field
BFETS Phase 3a	Natural gas-air mixtures	Complex Geometry	Near field
BFETS Phase 3b	Natural gas-air mixtures	Complex Geometry	Near field
DNVGL Explosion Chamber	Natural gas-air mixtures	Confined	Near field / Far field
MERGE	Methane, Propane	Confined	Near field / Far field

3.1 Quantitative Model Performance Assessment

The quantitative evaluation of a model is aimed at obtaining numerical data from the comparison of model predictions with experimental measurements, which can be used to whether a model is acceptable for hazard modeling, and whether a validation safety factor should be applied to the modeling results. Two different approaches have been used for the quantitative evaluation of a model:

1. by defining a set of statistical performance measures (SPMs) that compare predicted and observed physical comparison parameters;
2. by establishing a maximum probability of exceedance for the model's prediction of a critical parameter's value.

3.1.1 Statistical Performance Measures

The SPMs should be obtained by comparing model predictions to experimental measurements. Duijm et al. [9] suggested that a suitable set of SPMs should:

- Provide an indication of the model's bias (i.e., whether on average it under- or over-predicts the measurements)
- Provide an indication of the level of scatter (i.e., deviation from the average)
- Give equal weight to all measurements/predictions, regardless of their absolute values

Chang and Hanna [10] suggested that multiple statistical performance measures should be considered when evaluating a model, as each measure has its advantages and disadvantages. The statistical performance measures selected to evaluate vapor cloud explosion models are listed in Table 2.

Table 2. Statistical Performance Measures for Vapor Cloud Explosion.

SPM	Definition
Mean Relative Bias	$MRB = \langle \frac{c_m - c_p}{\frac{1}{2}(c_m + c_p)} \rangle$
Mean Relative Square Error	$MRSE = \langle \frac{(c_m - c_p)^2}{\frac{1}{4}(c_m + c_p)^2} \rangle$
Fraction of Predictions Within a Factor of 2	$0.5 \leq \left(\frac{c_p}{c_m}\right) \leq 2.0$
Geometric Mean Bias	$GMB = e^{\langle \ln\left(\frac{c_m}{c_p}\right) \rangle}$
Geometric Variance	$GV = e^{\langle \left(\ln\left(\frac{c_m}{c_p}\right)\right)^2 \rangle}$

SPMs should be calculated separately for each available trial, as well as combined for groups of trials with similar characteristics; they should then be compared to predefined acceptability criteria. In previous MVDs, the acceptability range for each SPM was set according to earlier model evaluation efforts, largely to provide continuity and consistency with the model evaluation protocols previously used by PHMSA to evaluate dispersion models [11]. However, at the basis of that approach is an assumption of what constitutes a "good" model, whose performance is then used as a comparison to determine whether other models are also "good" or not. In the case of VCE models, no acceptability criteria have ever been proposed, therefore, it is expected that the first few VCE models to be evaluated under this protocol will have to be assessed by a variety of means, to determine their acceptability for LNG facility siting. Once one or (preferably) more models have been accepted, their SPMs can be used to define an initial set of SPM acceptability criteria.

3.1.2 Maximum Exceedance Probability

The second approach to quantitative analysis of a model's performance is based on the calculation of bias and scatter of the model predictions relative to selected experimental data points. This method [12] considers that there are three sources of uncertainty:

- Experimental uncertainty, which is the uncertainty in experimental measurements of the comparison variable. This is often reported in test series summaries, or can be estimated based on similar instruments used in other test series.

- Propagated input uncertainty, which is the uncertainty in model parameter inputs (e.g., vapor cloud concentration, wind speed, etc.) due to uncertainty in experimental measurements of those parameters. It is calculated based on the change in model predictions when the input parameters are varied over the range of measurement uncertainty.
- Model uncertainty, which is the uncertainty arising directly from the model. It can be calculated based on the other two uncertainties and the model's predictions relative to a series of measurements.

A comparison of model predictions to experimental measurements allows the model bias and scatter to be calculated, which are then used to calculate the probability that a model prediction will exceed a given target value; this probability is compared with a predefined acceptability criterion and a validation factor is calculated which, if applied to the model predictions, would meet the acceptability target.

In addition to these two quantitative assessment approaches, the MVD spreadsheet produces plots which compare the model predictions to the experimental measurements for each trial in the database, as well as for groups of similar trials. A review of the plotted data allows a qualitative, yet helpful assessment of any modeling trends – for example, a model could have a very small overall bias either because it tracks the experimental data well across the entire range of measurements, or because it overpredicts in one area and underpredicts in another.

4 Summary

This paper described the scope of the Model Evaluation Protocol for vapor cloud explosions models; this is the fourth in a series of five protocols aimed at providing PHMSA, as well as other regulatory bodies that may choose to follow the same approach, with the tools to evaluate models for the different hazards associated with the siting of LNG facilities.

The paper provided a summary of the main sections of the MEP report, as well as a list of the test series included in the VCE model validation database. Readers interested in additional information were directed to a publicly accessible website which hosts the MEP and MVD reports.

The paper also discussed the quantitative assessment of model performance, which combines the "traditional" approach based on statistical performance measures (e.g., mean relative bias, geometric variance, etc.) and a more recent approach based on maximum probability of exceedance.

5 References

[1] O. R. Hansen, F. Gavelli, S. G. Davis, and P. Middha, "Equivalent cloud methods used for explosion risk and consequence studies," presented at the 14th Annual Symposium, Mary Kay O'Connor Process Safety Center, Oct. 2011.

[2] A. Beccantini, A. Malczynski, and E. Studer, "COMPARISON OF TNT-EQUIVALENCE APPROACH, TNO MULTI-ENERGY APPROACH AND A CFD APPROACH IN

INVESTIGATING HEMISPHERIC HYDROGEN-AIR VAPOR CLOUD EXPLOSIONS.," in *Proceedings of the 5th International Seminar on Fire and Explosion Hazards,* 2007, p. 18.

[3] M. J. Tang and Q. A. Baker, "A new set of blast curves from vapor cloud explosion," *Process Saf. Prog.*, vol. 18, no. 4, pp. 235–240, 1999, doi: 10.1002/prs.680180412.

[4] A. J. Pierorazio, J. K. Thomas, Q. A. Baker, and D. E. Ketchum, "An update to the Baker-Strehlow-Tang vapor cloud explosion prediction methodology flame speed table," *Process Saf. Prog.*, vol. 24, no. 1, pp. 59–65, Mar. 2005, doi: 10.1002/prs.10048.

[5] F. Gavelli, B. Hendrickson, J. Kolade, and R. Kooy, "Towards a comprehensive model evaluation protocol for LNG hazard analyses," presented at the 23rd Annual Process Safety International Symposium, College Station, TX, Oct. 2020, p. 14.

[6] T. Skjold *et al.*, "A Matter of Life and Death: Validating, Qualifying and Documenting Models for Simulating Flow-Related Accident Scenarios in the Process Industry," *Chem. Eng. Trans.*, vol. 31, 2013.

[7] F. Gavelli, B. Hendrickson, C. Marsegan, J. Piekarz, R. Kooy, and B. Kolade, "Model Evaluation Protocol for Flammable Dispersion," Pipeline and Hazardous Materials Safety Administration, 03903-RP-002, Oct. 2020.

[8] F. Gavelli, B. Hendrickson, C. Marsegan, J. Piekarz, R. Kooy, and B. Kolade, "Model Evaluation Protocol for Toxic Dispersion," Pipeline and Hazardous Materials Safety Administration, 03903-RP-005, Dec. 2020.

[9] N. J. Duijm, S. Ott, and M. Nielsen, "An evaluation of validation procedures and test parameters for dense gas modeling," *J. Loss Prev. Process Ind.*, vol. 9, no. 5, pp. 323–338, 1996.

[10] J. C. Chang and S. R. Hanna, "Air quality model performance evaluation," *Meteorol. Atmospheric Phys.*, vol. 87, no. 1–3, Sep. 2004, doi: 10.1007/s00703-003-0070-7.

[11] M. J. Ivings, S. F. Jagger, C. J. Lea, and D. M. Webber, "Evaluating vapor dispersion models for safety analysis of LNG facilities," Health and Safety Laboratory, MSU/2007/04, 2007.

[12] A. Luketa, "Guidance on the Evaluation of Models for LNG fires," SAND2019-4426, Apr. 2019.

51b Risk Assessment of Hydrogen for Use in Hazardous Process

Abhineet Raj, Naveen Kumar

Background:

CO2 emissions in BF-BOF based steel industry is mainly attributed to usage of coal as a fuel & reducing agent for smelting of iron ore. This results in emission of approximately 1.80 tons of CO2 per tonne of hot metal. In line with fuel rate reduction in iron making through blast furnace route, alternate fuel injection, alternate reductants are being practiced in different parts of the world. Most of these also are hydrogen rich. These injectants also lead to reduction of CO2 footprint, thereby participating in carbon direct avoidance (CDA). Different such possible injectants, that are being in use are oils, tar, natural gas, hydrogen, coke oven gas etc.

Our company recently concluded its one of its kind, month-long trial with Coal Bed Methane (CBM) injection in one of its blast furnaces with encouraging results in terms of lowering of coke rates. Considering the same, effort was made to inject pure hydrogen in one of its blast furnaces.

News publication link:

Tata Steel initiates trial for record-high hydrogen gas injection in Blast Furnace at its Jamshedpur Works

Scope of the project:

Hydrogen injection trial was done in 4 tuyeres via hydrogen tanker. This helped in gaining important answers related to technology required for safe handling and injection of hydrogen in blast furnace.

Project Description:

Trial was carried out with hydrogen injection via 4 tuyeres in one of the blast furnaces. Main objectives of this revolves around learning usage of hydrogen and are listed as follows:

- Hydrogen handling and safety concerns
- Basic engineering
- Plant technology requirements
- Suitability and compatibility of injection hardware

Hydrogen supply for these trials was sourced using truck mounted cascades from potential suppliers along with necessary skid for regulating pressure for injection. Pressure reduction skid and transportation pipelines was laid from tanker position till designated tuyere for injection with all safety features and interlocks.

Injection Scheme:

In the existing tar injection system at one of the blast furnaces, there is a compressed air system present for cooling of the lance. In the concentric lance, tar travels from the center and compressed air flows through the annulus. In this system, compressed air system was used for hydrogen injection with no-tar.

The proposed system was itself a unique design and was modified considering the hazards of blast furnace and peripheral area

Challenges in Hydrogen Handling:

- Very Low Ignition energy - 0.017 mJ as compared to other gases.
- Wide range of explosive mixture 4 – 72% as compared to other gases.
- Leakage probability due to low molecular weight of H2 & High pressure handling
- Colourless and odourless and can embrittle some metals

Uniqueness:-

- HMI developed with PSM critical real time parameters with more than 100 safety interlock was built in the logic.
- Automatic stoppage of H2 in case of trip activation interlock and immediate switchover to N2 Inertising/ purging sequence in each tripping.
- Real time simulations & testing of all safety interlocks was done before injection start up.
- Total no of H2 Tankers used –25 Nos with Hydrogen Tanker Pressure - 200 bar and Furnace Injection pressure – 4.5 Bar
- Two stage Pressure reduction from 200 bar to 8 bar and from 8 bar to 4.5 bar with pipeline distance from tanker to furnace ~ 300 mtr
- Excess flow check valve and three way valve with changeover facility and start of purging and venting
- Elimination of entire electrical system from high pressure PRS where multi fold pressure reduction (15 times) is done
- The 1st stage PRS is an inherently safer and reliable design as it consists of mechanical system only, even the flow control valves etc are free from electronic and electrics.
- Safe logistic management of tankers to support the peak hydrogen injection rate @ above 1500 nm3/hr
- The entire injection process of 38000Nm3 is completed safely without a single leak
- Unique system followed to detect leak during injection trial of duration 60 hrs with the help of portable electrochemical H2 PPM detector and Thermography camera as hydrogen leak phenomenon follows opposite to Joule Thomson effect
- There was no unintended leakage of hydrogen.
- Hazop, Consequence modelling through PHAST software, Bow Tie, PSSR conducted before the trial.

Achievements:

- Our company became 1st to inject such large volume of hydrogen in any Blast Furnace
- During Hazop, 52 number of high risk scenarios were identified and mitigated at the design phase itself
- Hydrogen Injection is one of the key strategy of our company in the decarbonisation journey . The endeavour is aligned with the Company's vision of becoming Net Zero by 2045 kick started with this project
- The trial has the potential to reduce the coke rate by 10%, translating into around 7-10% reduction in CO2 emissions per ton of crude steel produced saving millions of money.
- Very encouraging process results were observed during the trial and team learnt Safe handling/managing hydrogen injection in Blast Furnace.
- The process sees hydrogen substituted in for carbon as the reducing agent in the extraction process at 40% of the furnace's injection systems.

51c Defining Grid Resolution for Trench Fire Modeling with FDS

Thomas Butts, Bryant Hendrickson, Filippo Gavelli

Trench fires are often the design basis for fireproofing of piperacks and other critical components in a hydrocarbon processing facility. Prescriptive guidance based on API 2218 Fireproofing Practices in Petroleum and Petrochemical Processing Plants generally results in fireproofing being applied within 30-ft of a trench. Alternatively, fire modeling can be conducted as part of a performance-based approach to determine the extent of fireproofing necessary. Trench fires pose a unique challenge for modeling due to the smaller dimensions of the trench relative to other pool fire scenarios (e.g., sump fires). Current guidelines for fire models used at LNG facilities recommend at least 20 cells across the surface of a fire. However, this often leads to very high cell counts and simulations that require impractical computational resources and time.

This paper will detail a grid size sensitivity analysis for modeling trench fires with Fire Dynamics Simulator (FDS). By examining various trench fire scenarios (i.e., altering grid resolution for FDS validation cases) we aim to establish guidelines for more practical grid sizes. Balancing the need for accuracy and computational efficiency, our research contributes to the reliability of CFD models in the assessment of trench fire impacts on LNG facility design.

138a Engineering Experimental Approach to Study Methane Dehydroaromatization Stability to CO2 in the Feed Stream

Maria Haki, Patrick Linke, Izabel Medeiros-Costa, Mamoun Al-Rawashdeh

Methane, is the major constituent of natural gas that can be used to produce a wide range of valuable products. Methane can be converted using one of two routes: the indirect and direct route. The indirect route is the commercialized one that involves producing syngas as an intermediate and converting it to other chemicals. The direct route includes catalytic reactions that convert methane to products in a single step with lower CO_2 emissions and simplified processes. One of the major direct reactions that has been studied extensively recently, is the methane dehydroaromatization (MDA) that produces aromatics, mainly, benzene and hydrogen. This reaction is challenged by the low thermodynamic conversion, need for high temperature, and fast catalyst deactivation. These challenges have been investigated by many researchers. One technical aspect that has not been investigated much yet is the MDA catalyst and process stability to CO_2. Even though the methane coming from its source natural gas passes through different treatment processes, CO_2 and other impurities can still be present in the feed stream. Besides, having an MDA catalyst that is stable to CO_2 opens the possibility for chemical and thermal synergies by coupling the MDA with a second reaction to overcome some of the technical challenges of the MDA reaction. In the literature, some studies considered the addition of different CO_2 amounts to the MDA feed and analyzed its impact [1]. Other studies attempt to co-feed CO_2 and other products from a second reaction [2]. Furthermore, some efforts were put into adding a second metal to the Mo catalyst to prevent the active-inactive catalyst conversion. For this purpose, different metals have been studied including Ga [3], Fe [4], K [4], and Mg [5].

This work wants to investigate the possibility of enhancing the resistivity of MDA to CO_2 through several experimental studies, as this is not readily available to investigate this effect theoretically at the moment. First, CO_2 was fed to the MDA catalyst (6wt% Mo-ZSM5) in different amounts and modes of addition using pulse and periodic feeding. These feeding modes were valuable in evaluating the MDA catalyst stability to CO_2. A testing protocol was developed to screen other MDA catalysts. This included potassium-promoted k-Mo/ZSM-5 and gallium-based ZSM-5 catalyst. Afterward, an experimental setup that contained two separate fixed bed reactors in series was utilized. The second reactor contained the MDA catalyst while the first one contained another catalyst that can process methane to generate CO_2 and other components. This approach allowed us to investigate more realistic feedstock to the MDA catalyst and explore the possibility of coupling the MDA catalyst to another reaction. The obtained findings from all these experimental studies will be presented to provide directions and better understanding the impact of CO_2 on the MDA catalyst and process development.

References

[1] Liu, S., Wang, L., Ohnishi, R., & Lchikawa, M. (2000). Bifunctional catalysis of Mo/HZSM-5 in the dehydroaromatization of methane with CO/CO2 to benzene and naphthalene. *Kinetics and Catalysis*, *41*(1), 132–144. https://doi.org/10.1007/bf02756152

[2] Li, Y., Liu, H., Shen, W., Bao, X., & Xu, Y. (2004). Combined single-pass conversion of methane via oxidative coupling and dehydroaromatization: a combination of La2O3/BaO and Mo/HZSM-5

catalysts. In *Studies in Surface Science and Catalysis* (pp. 583–588). https://doi.org/10.1016/s0167-2991(04)80115-0

[3] Liu, B., Yang, Y., & Sayari, A. (2001). Non-oxidative dehydroaromatization of methane over Ga-promoted Mo/HZSM-5-based catalysts. *Applied Catalysis A-General*, *214*(1), 95–102. https://doi.org/10.1016/s0926-860x(01)00470-7

[4] Ramasubramanian, V., Lienhard, D. J., Ramsurn, H., & Price, G. L. (2019). Effect of addition of K, RH and FE over MO/HZSM-5 on methane dehydroaromatization under non-oxidative conditions. *Catalysis Letters*, *149*(4), 950–964. https://doi.org/10.1007/s10562-019-02697-8

[5] Ryu, H. W., Lim, Y. H., Jung, W., Nam, K., Hwang, Y., Roh, J. E., & Kim, D. H. (2023). Enhanced stability of CO2-assisted shale gas aromatization through the introduction of Mg promoter on Mo/ZSM-5. *Chemical Engineering Journal*, *467*, 143404. https://doi.org/10.1016/j.cej.2023.143404

138c Multi-Objective Bayesian Optimization of NGL Fractionators Using Process Simulation

Clinton Duewall, Mahmoud El-Halwagi

After remaining relatively constant for over 30 years, domestic natural gas liquids production has tripled since 2010 as a direct result of the U.S. Shale Revolution. This work presents an approach to optimize natural gas liquids fractionator throughput from existing plants using multi-objective optimization. The method maximizes operating net revenue while simultaneously minimizing direct and indirect emission rates by applying a custom multi-objective variation of the efficient global optimization (EGO) algorithm. This algorithm optimizes kriging surrogate models of the objective functions to determine which points to sample in the process simulation before the next iteration. The Pareto front obtained from multi-objective EGO is compared with results from other optimization methods to recommend both a formulation and a solution method to economically and environmentally optimize a natural gas liquids fractionator.

98a Affordable "Green-ish' Production of Hydrogen from Methane

Eric McFarland

A rapid transition from a world powered primarily by low-cost fossil fuel combustion to more sustainable low emissions alternatives will only occur if those alternatives are widely available at comparable costs. In addition to electricity, hydrogen and other hydrogen containing fuels (e.g. NH_3) will likely be important in future sustainable energy economies provided their production costs are low. Hydrogen is produced today, for profit, from abundant low-cost methane by reforming and complete oxidation of the carbon from C(-4) to C(+4) producing ~ 7 tons CO_2/ton H_2. Partial oxidation of methane and other fossil resources in the absence of oxygen can produce solid carbon C(0) and potentially valuable hydrogen containing molecules without generating CO_2 at far lower cost than producing hydrogen from water by electrolysis. If society is willing to pay for CO_2 emissions reductions, methane pyrolysis can be the lowest cost option for producing hydrogen as a fuel. Methane pyrolysis is equilibrium limited and thermodynamics restrict high pressure industrial processes to high temperatures for acceptable conversion. Catalysis can enable the reaction rates and conditions to be practical at commercial scales. The evolution of several catalyst-based processes will be discussed, and their relative merits compared for producing hydrogen at large-scale without carbon dioxide emissions.

98b Potential for Decarbonizing Ammonia Production Using Cargen® Technology

Nada Mahmoud, Mohamed Sufiyan Challiwala, Nimir Elbashir

Ammonia (NH_3) stands as a pivotal industrial chemical, primarily synthesized through the renowned Haber-Bosch process. Beyond its significance in agriculture, ammonia has garnered attention as a promising energy carrier due to its substantial hydrogen (H_2) content. However, ammonia production demands considerable energy, resulting in approximately 2% of the world's total energy consumption. Consequently, ammonia synthesis contributes up to 1.6% of global CO_2 emissions and 5% of CO_2 emissions from chemical industries. In particular, 1.25-2.16 kg of CO_2 is emitted per kg of ammonia produced in conventional ammonia production. Therefore, extensive efforts are underway to decarbonize the process for ammonia production to become environmentally sustainable. These efforts include transforming existing coal and natural gas-based ammonia plants (often termed black or grey ammonia) into blue ammonia facilities by integrating carbon capture and sequestration (CCS) technology or transitioning to green ammonia plants powered by renewable energy sources. However, green ammonia production, in particular, faces challenges in scalability due to the decentralized and sporadic availability of renewable energy and the elevated costs associated with it, which diminishes the economic viability of green ammonia.

One approach to significantly reducing CO_2 emissions in ammonia synthesis involves the application of innovative CARGEN® (CARbon GENerator) technology, which converts CO_2 and methane into multi-walled carbon nanotubes (MWCNTs) and hydrogen-rich syngas. The Texas A&M University technology has successfully reduced the energy consumption of the conventional methane dry reforming (DRM) process by 50% and demonstrated a remarkable reduction in CO_2 emissions by up to 73% upon integration in conventional gas-to-liquid (GTL) process plants. Therefore, the work aims to retrofit ammonia reformers with Carbon Capture and Utilization (CCU) systems such as CARGEN® technology to decarbonize conventional ammonia production sustainably and economically.

In the present study, a conventional ammonia synthesis process flowsheet with a capacity of 1,265 tons per day has been developed to identify various sources of direct CO_2 emissions and assess energy and power utilization. The simulation has been developed using the ASPEN® Plus process simulator and serves as a reference point for comparison with a similarly sized ammonia plant employing CARGEN® technology reformer. The retrofitting approach replaced the combination of steam methane reformer (SMR) and autothermal reformer (ATR) processes with an equivalent-capacity CARGEN® process that yields a product stream composition similar to that of the base case reformer unit. Next, a comprehensive mass and energy balance analysis was conducted to evaluate the CO_2 utilization and abatement potential of the CARGEN®-based ammonia plant compared to the base case ammonia process. Our preliminary results demonstrate that besides full CO_2 recycling from

the amine separation unit, the CARGEN® process consumes a surplus of 3,000 tons/day of CO_2 to fulfill its CO_2 feedstock requirements.

Moreover, the CARGEN® based ammonia process requires up to 67% less steam than the base case process; however, the process requires about 2.5 times more natural gas compared to the base case process. In addition, the results show that ammonia plants based on the CARGEN® process have 19% higher energy requirements and 11% higher power requirements due to additional supply of natural gas and CO_2. However, plant-wide consideration of direct and in-direct CO_2 emissions reveals that the CARGEN® Technology integration reduces the net CO_2 emission from 3,013 to 1,321 tons/day. Consequently, an ammonia plant based on CARGEN® Technology reduces the overall carbon footprint of the process to 2.16 kg CO_2-eq/kg NH_3 compared to 4.76 kg CO_2-eq/kg NH_3 in the conventional process.

This paper will provide more details on the potential impact of the CARGEN® technology for decarbonizing ammonia plants to provide a better carbon footprint product than blue ammonia. This innovative approach offers a promising pathway for generating carbon-negative ammonia, presenting a significant and alternative technology to address the challenges posed by climate change.

References:

1. Ataya, Zeinab., et al. 'Decarbonizing the Gas to Liquid (GTL) process using an advanced reforming of methane process' ACS Engineering Au, DOI 10.1021/acsengineeringau.3c00025

2. Challiwala, Mohamed S., et al. 'A novel CO2 utilization technology for the synergistic co-production of multi-walled carbon nanotubes and syngas.' Scientific reports 11.1 (2021): 1417.

3. Challiwala, Mohamed S., et al. 'Alternative Pathways for CO2 Utilization via Dry Reforming of Methane.' Advances in Carbon Management Technologies. CRC Press, 2020. 253-272.

4. Challiwala, M. S., et al. 'A combined thermo-kinetic analysis of various methane reforming technologies: Comparison with dry reforming.' Journal of CO2 Utilization 17 (2017): 99-111.

5. Challiwala, Mohamed S., et al. 'System and method for Carbon and Syngas Production'. US11591213B2; AU2018249486B2; PCT/US2018/025696; WO2018187213

AUTHOR INDEX

AIChE
120 Wall Street, FL 23
New York, NY 10005-4020

ISBN 979-8-3313-1708-9

CURRAN ASSOCIATES INC.
proceedings.com